The New Wind Has Wings

Poems
from
Canada

Compiled by Mary Alice Downie & Barbara Robertson

Illustrated by Elizabeth Cleaver

Toronto Oxford
Oxford University Press
1984

Contents

CANADIAN CATALOGUING IN PUBLICATION DATA
Main entry under title:
The New wind has wings
Previous ed. had title: The Wind has wings.
Includes index.
ISBN 0-19-540431-9 (bound). — ISBN 0-19-540432-7 (pbk.)

1. Canadian poetry. 2. Children's poetry, Canadian.
I. Downie, Mary Alice, 1934–   II. Robertson,
Barbara, 1931–   III. Cleaver, Elizabeth, 1939–
IV. Title: The Wind has wings.

PS8273.W56 1984   C811'.008   C84-098325-5
PR9195.25.W56 1984

OXFORD is a trademark of Oxford University Press
Selection © Oxford University Press Canada 1984
1 2 3 — 6 5 4   Printed in Hong Kong

Orders

Muffle the wind;
Silence the clock;
Muzzle the mice;
Curb the small talk;
Cure the hinge-squeak;
Banish the thunder.
Let me sit silent,
Let me wonder.

A. M. KLEIN

5

Rhymes

Two respectable rhymes
skipped out of their pages
like two proud roosters
from golden cages;

they walked many a mile
in search of a home,
but could find no space
for themselves in a poem.

They grew tired and sad
but wherever they went
nobody advertised
poems for rent.

People whispered and said:
haven't you heard
that a rhyming word
is considered absurd?

In modern times
who needs rhymes?
Those high-flying words
went out with the birds.

At last one night
all weary and worn
they came to a house
in a field of corn;

and there lived a man
who still wrote lines
according to rules
from olden times.

So he took them in
with doubles and pairs,
and set them to music,
and gave them new airs.

Now they ring again
their bells and chimes,
and the children all sing
those respectable rhymes,

with one rhyme inside
and another one out:
the rhymes were befriended
and my poem is ended.

Y. Y. SEGAL
Translated from the Yiddish
by MIRIAM WADDINGTON

King Rufus

To a king who had
A red little beard,
A red little horse
Was most endeared.

And also red
Were his jerkin and hose;
The monarch loved
A red, red rose.

And crimson apples
The monarch sought
For cooking in
His scarlet pot.

In fine, a royal
Person warm in
Vestments that were
Wholly carmine.

When one beheld him
Journey his course
Upon his fleet little
Russet horse

He seemed to crackle
He seemed to be
An ever burning
Granary . . .

Scarlet, rufous,
Roseate,
What a fiery
Potentate!

Y. Y. SEGAL
*Translated from the Yiddish
by* A. M. KLEIN

9

Laughter

We are light
as dandelion
parachutes we
land anywhere
take the shape
of wherever we
fall

we are often
the size of
grasshoppers in
a jungle of grass
or we're squirmy
chains of willow
catkins

then we become
curly seashells
knobby little
swimmers in a
sea of air
lying

on our backs
our eyes fly up
higher than kites
airplanes clouds
winds higher
than stars and

we stare down
at the little
distant world
and we laugh
laugh laugh

MIRIAM WADDINGTON

The Boar and the Dromedar

A wealthy dromedar
sat smoking a cigar
in a quiet desert bar,
when in came with a roar
a mighty tourist boar
who almost broke the door.
He ordered ginger-beer
and for a souvenir
wanted the chandelier.
Up rose the dromedar,
put out his big cigar
and trotted to the bar:
"Wild pig, get out of here
or else I'll box your ear!
Myself will take the chandelier!"
He called a taxi-star,
went straight to a bazaar,
and for a big cigar
exchanged the chandelier.
O dear, O dear, O dear!
The boar in his despair
became a whiff of air
shaped like a prickly pear,
and rolled back out the door.
Since then no desert bar
will serve a passing boar
even ginger-beer.

HENRY BEISSEL

The Ships of Yule

When I was just a little boy,
Before I went to school,
I had a fleet of forty sail
I called the Ships of Yule;

Of every rig, from rakish brig
And gallant barkentine,
To little Fundy fishing boats
With gunwales painted green.

They used to go on trading trips
Around the world for me,
For though I had to stay on shore
My heart was on the sea.

They stopped at every port to call
From Babylon to Rome,
To load with all the lovely things
We never had at home;

With elephants and ivory
Bought from the King of Tyre,
And shells and silks and sandal-wood
That sailor men admire;

With figs and dates from Samarcand,
And squatty ginger-jars,
And scented silver amulets
From Indian bazaars;

With sugar-cane from Port of Spain,
And monkeys from Ceylon,
And paper lanterns from Pekin
With painted dragons on;

With cocoanuts from Zanzibar,
And pines from Singapore;
And when they had unloaded these
They could go back for more.

And even after I was big
And had to go to school,
My mind was often far away
Aboard the Ships of Yule.

BLISS CARMAN

Jack Was Every Inch a Sailor

Now, 'twas twenty-five or thirty years since Jack first saw the light;
He came into this world of woe one dark and stormy night.
He was born on board his father's ship as she was lying to
'Bout twenty-five or thirty miles south-east of Bacalhao.

> CHORUS
> *Jack was ev'ry inch a sailor,*
> *Five and twenty years a whaler;*
> *Jack was ev'ry inch a sailor,*
> *He was born upon the bright blue sea.*

When Jack grew up to be a man, he went to Labrador;
He fished in Indian Harbour where his father fished before;
On his returning in the fog, he met a heavy gale,
And Jack was swept into the sea and swallowed by a whale.

The whale went straight for Baffin's Bay 'bout ninety knots an
    hour,
And ev'ry time he'd blow a spray, he'd send it in a shower.
"Oh, now," says Jack unto himself, "I must see what he's about."
He caught the whale all by the tail and turned him inside out.

ANONYMOUS

14

Donkey Riding

Were you ever in Quebec
Stowing timber on the deck,
Where there's a king with a golden crown,
   Riding on a donkey?

CHORUS
Hey, ho! Away we go!
Donkey riding, donkey riding,
Hey, ho! Away we go!
Riding on a donkey.

Were you ever off the Horn
Where it's always fine and warm,
Seeing the Lion and the Unicorn,
   Riding on a donkey?

Were you ever in Cardiff Bay
Where the folks all shout "Hooray!
Here comes John with his three months' pay",
Riding on a donkey?

ANONYMOUS

"So They Went Deeper into the Forest . . ."

"So they went deeper into the forest," said Jacob Grimm,
And the child sat listening with all his ears,
While the angry queen passed. And in after years
The voice and the fall of words came back to him
(Though the fish and the faithful servant were grown dim,
The aproned witch, the door that disappears,
The lovely maid weeping delicious tears
And the youngest brother, with one bright-feathered limb) —
"Deeper into the forest."
                                        There are oaks and beeches
And green high hollies. The multitudinous tree
Stands on the hill and clothes the valley, reaches
Over long lands, down to a roaring sea.
And the child moved onward, into the heart of the wood,
Unhindered, unresisted, unwithstood.

ROY DANIELLS

Eating Fish

Here is how I eat a fish
    —Boiled, baked or fried—
Separate him in the dish,
    Put his bones aside.

Lemon juice and chive enough
    Just to give him grace,
Make of his peculiar stuff
    My peculiar race.

Through the Travellers' Hotel
    From the sizzling pan
Comes the ancient fishy smell
    Permeating man.

May he be a cannier chap
    Altered into me,
Eye the squirming hook, and trap,
    Choose the squirming sea.

GEORGE JOHNSTON

Egg

Reader, in your hand you hold
A silver case, a box of gold.
I have no door, however small,
Unless you pierce my tender wall,
And there's no skill in healing then
Shall ever make me whole again.
Show pity, Reader, for my plight:
Let be, or else consume me quite.

JAY MACPHERSON

O Earth, Turn!

The little blessed Earth that turns
Does so on its own concerns
As though it weren't my home at all;
It turns me winter, summer, fall
Without a thought of me.

I love the slightly flattened sphere,
Its restless, wrinkled crust's my here,
Its slightly wobbling spin's my now
But not my why and not my how:
My why and how are me.

GEORGE JOHNSTON

The Reformed Pirate

His proper name was Peter Sweet:
But he was known as Keel-haul Pete
From Turtle Cay to Port-of-Spain
And all along the Spanish Main,
And up and down those spicy seas
Which lave the bosky Caribbees.
His sense of humour was so grim,
Fresh corpses were but jokes to him.
He chuckled, chortled, slapt his flank,
To see his victims walk the plank.
His language — verbal bilge and slush —
Made all who heard it quake and blush.
Loud would he laugh, with raucous jeers,
To see his shipmates plug their ears
Whenever, feeling extra gay,
To his high spirits he gave way.

But were his shipmates prudes? Oh no! —
Ptomaine Bill and Strangler Joe,
Slicer Mike, Tarnation Shay,
And twoscore more as bad as they,
Ready to cut throats any day.
But Pete's expressions used to freeze
E'en their tough sensibilities.
Like shocked young ladies they would cry,
"Avast!" "Belay!" and "Fie, oh fie!"
Pete's home-life was not — well, quite nice.
In one short week he married thrice;
And so on. All his cool retreats
(From which had fled the parakeets)
Were over-run with Missus Sweets:
And yet his heart was ever true —
Deep down — to Angostura Sue.

Three nights hand-running — one, two, three —
He dreamed about a gallows-tree.
Three nights hand-running, he awoke
With yells that made the bulkheads smoke.
Then terror took his soul by storm:
So he decided to reform.

T. G. ROBERTS

21

The Shooting of Dan McGrew

A bunch of the boys were whooping it up in the Malamute saloon;
The kid that handles the music-box was hitting a jag-time tune;
Back of the bar, in a solo game, sat Dangerous Dan McGrew,
And watching his luck was his light-o'-love, the lady that's known as Lou.

When out of the night, which was fifty below, and into the din and the glare,
There stumbled a miner fresh from the creeks, dog-dirty, and loaded for bear.
He looked like a man with a foot in the grave and scarcely the strength of a louse,
Yet he tilted a poke of dust on the bar, and he called for drinks for the house.
There was none could place the stranger's face, though we searched ourselves for a clue;
But we drank his health, and the last to drink was Dangerous Dan McGrew.

There's men that somehow just grip your eyes, and hold them hard like a spell;
And such was he, and he looked to me like a man who had lived in hell;
With a face most hair, and the dreary stare of a dog whose day is done,
As he watered the green stuff in his glass, and the drops fell one by one.
Then I got to figgering who he was, and wondering what he'd do,
And I turned my head — and there watching him was the lady that's known as Lou.

His eyes went rubbering round the room, and he seemed in a kind of daze,
Till at last that old piano fell in the way of his wandering gaze.
The rag-time kid was having a drink; there was no one else on the stool,
So the stranger stumbles across the room, and flops down there like a fool.
In a buckskin shirt that was glazed with dirt he sat, and I saw him sway;
Then he clutched the keys with his talon hands — my God! but that man could play!

Were you ever out in the Great Alone, when the moon was awful clear,
And the icy mountains hemmed you in with a silence you most could *hear;*
With only the howl of a timber wolf, and you camped there in the cold,
A half-dead thing in a stark, dead world, clean mad for the muck called gold;
While high overhead, green, yellow and red, the North Lights swept in bars? —
Then you've a hunch what the music meant . . . hunger and night and the stars.

And hunger not of the belly kind, that's banished with bacon and beans,
But the gnawing hunger of lonely men for a home and all that it means;
For a fireside far from the cares that are, four walls and a roof above;
But oh! so cramful of cosy joy, and crowned with a woman's love —
A woman dearer than all the world, and true as Heaven is true —
(God! how ghastly she looks through her rouge, — the lady that's known as Lou.)

Then on a sudden the music changed, so soft that you scarce could hear;
But you felt that your life had been looted clean of all that it once held dear;
That someone had stolen the woman you loved; that her love was a devil's lie;
That your guts were gone, and the best for you was to crawl away and die.
'Twas the crowning cry of a heart's despair, and it thrilled you through and through —
"I guess I'll make it a spread misere," said Dangerous Dan McGrew.

The music almost died away . . . then it burst like a pent-up flood;
And it seemed to say, "Repay, repay," and my eyes were blind with blood.
The thought came back of an ancient wrong, and it stung like a frozen lash,
And the lust awoke to kill, to kill . . . then the music stopped with a crash,
And the stranger turned, and his eyes they burned in a most peculiar way;
In a buckskin shirt that was glazed with dirt he sat, and I saw him sway;
Then his lips went in in a kind of grin, and he spoke, and his voice was calm,
And, "Boys," says he, "you don't know me, and none of you care a damn;
But I want to state, and my words are straight, and I'll bet my poke they're true,
That one of you is a hound of hell . . . and that one is Dan McGrew."

Then I ducked my head, and the lights went out, and two guns blazed in the dark,
And a woman screamed, and the lights went up, and two men lay stiff and stark.
Pitched on his head, and pumped full of lead, was Dangerous Dan McGrew,
While the man from the creeks lay clutched to the breast of the lady that's known as Lou.

These are the simple facts of the case, and I guess I ought to know.
They say that the stranger was crazed with "hooch", and I'm not denying it's so.
I'm not so wise as the lawyer guys, but strictly between us two —
The woman that kissed him and — pinched his poke — was the lady that's known as Lou.

ROBERT W. SERVICE

The Yak

For hours the princess would not play or sleep
    Or take the air;
Her red mouth wore a look it meant to keep
    Unmelted there;
(Each tired courtier longed to shriek, or weep,
    But did not dare.)

Then one young duchess said: "I'll to the King,
    And short and flat
I'll say, 'Her Highness will not play or sing
    Or pet the cat;
Or feed the peacocks, or do anything —
    And that is that.' "

So to the King she went, curtsied, and said,
    (No whit confused):
"Your Majesty, I would go home! The court is dead.
    Have me excused;
The little princess still declines," — she tossed her head —
    "To be amused."

Then to the princess stalked the King: "What ho!" he roared,
    "What may you lack?
Why do you look, my love, so dull and bored
    With all this pack
Of minions?" She answered, while he waved his sword:
    "I want a yak."

"A yak!" he cried (each courtier cried, "Yak! Yak!"
    As at a blow)
"Is that a figure on the zodiac?
    Or horse? Or crow?"
The princess sadly said to him: "Alack
    I do not know."

"We'll send the vassals far and wide, my dear!"
    Then quoth the King:
"They'll make a hunt for it, then come back here
    And bring the thing; —
But warily, — lest it be wild, or queer,
    Or have a sting."

So off the vassals went, and well they sought
    On every track,
Till by and by in old Tibet they bought
    An ancient yak.
Yet when the princess saw it, she said naught
    But: "Take it back!"

And what the courtiers thought they did not say
    (Save soft and low),
For that is surely far the wisest way
    As we all know;
While for the princess? She went back to play!

      Tra-rill-a-la-lo!
      Tra-rill-a-la-lo!
      Tra-rill-a-la-lo!

VIRNA SHEARD

A Threnody

"The Ahkoond of Swat is dead." — PRESS DISPATCH

What, what, what,
What's the news from Swat?
    Sad news,
    Bad news,
Comes by the cable led
Through the Indian Ocean's bed,
Through the Persian Gulf, the Red
Sea and the Med-
Iterranean — he's dead;
The Ahkoond is dead!
For the Ahkoond I mourn.
    Who wouldn't?
He strove to disregard the message stern,
    But he Ahkoondn't.

26

Dead, dead, dead;
    Sorrow, Swats!
Swats wha' hae wi' Ahkoond bled,
Swats whom he had often led
Onward to a gory bed,
    Or to victory,
    As the case might be.
    Sorrow, Swats!
Tears shed,
    Shed tears like water,
Your great Ahkoond is dead!
    That Swat's the matter!

Mourn, city of Swat!
Your great Ahkoond is not,
But lain 'mid worms to rot:
His mortal part alone, his soul was caught
(Because he was a good Ahkoond)
Up to the bosom of Mahound.
Though earthly walls his frame surround
(For ever hallowed be the ground!)
And skeptics mock the lowly mound
And say, "He's now of no Ahkound!"
(His soul is in the skies!)

The azure skies that bend above his loved
    Metropolis of Swat
He sees with larger, other eyes,
Athwart all earthly mysteries —
    He knows what's Swat.

Let Swat bury the great Ahkoond
  With a noise of mourning and of lamentation!
Let Swat bury the great Ahkoond
  With the noise of the mourning of the Swattish nation!
  Fallen is at length
  Its tower of strength,
Its sun had dimmed ere it had nooned:
Dead lies the great Ahkoond.
  The great Ahkoond of Swat
  Is not.

GEORGE T. LANIGAN

The Gallant Highwayman

It was a gallant highwayman
    That stopped the Royal Mail;
The ladies shrieked and swooned away
The gentlemen turned pale.

"Forbear," the courteous robber said,
    "Your outcries and your curses,
For you can take your lives away
    By giving up your purses."

JAMES DE MILLE

Bandit

There was a Jewish bandit who lived in a wood,
He never did much evil, nor ever did much good,
For he would halt a merchant, quivering to his toes,
And in a gruff voice whisper: *You'll pay through the nose.*
And then he'd search his person, having bid him pray,
And snatch his broken snuff-box, and sneeze himself away.

A. M. KLEIN

Love Me, Love My Dog

He had a falcon on his wrist,
   A hound beside his knee,
A jewelled rapier at his thigh;
   Quoth he: "Which may she be?
My chieftain cried: 'Bear forth, my page,
   This ring to Lady Clare;
Thou'lt know her by her sunny eyes
   And golden lengths of hair.'
But here are lovely damsels three,
   In glittering coif and veil,
And all have sunny locks and eyes, —
   To which unfold the tale?"

Out spake the first: "O pretty page,
   Thou hast a wealthy lord;
I love to see the jewels rare
   Which deck thy slender sword!"
She smiled, she waved her yellow locks,
   Rich damask glowed her cheek;
He bent his supple knee and thought:
   "Not this the maid I seek."

The second had a cheek of rose,
   A throat as white as milk,
A jewelled tire upon her brow,
   A robe and veil of silk.
"O pretty page, hold back the hound;
   Uncouth is he and bold;
His rough caress will tear my veil,
   My fringe of glittering gold!"
She frowned, she pouted ruby lips —
   The page he did not speak;
He bent his curly head and thought:
   "Not this the maid I seek."

The third, with cobweb locks of light
    And cheeks like summer dawn,
Dropped on her knee beside the hound
    Upon the shaven lawn.
She kissed his sinewy throat, she stroked
    His bristly rings of hair;
"Ho!" thought the page, "she loves his hound,
    So this is Lady Clare!"

ISABELLA V. CRAWFORD

Sweet Maiden of Passamaquoddy

Sweet maiden of Passamaquoddy,
Shall we seek for communion of souls
Where the deep Mississippi meanders,
Or the distant Saskatchewan rolls?
Ah no! in New Brunswick we'll find it —
A sweetly sequestered nook —
Where the sweet gliding Skoodawabskooksis
Unites with the Skoodawabskook . . .

Let others sing loudly of Saco,
Of Passadumkeag or Miscouche,
Of the Kennebecasis or Quaco,
Of Miramichi or Buctouche;
Or boast of the Tobique or Mispec,
The Mushquash or dark Memramcook;
There's none like the Skoodawabskooksis
Excepting the Skoodawabskook.

JAMES DE MILLE

Isabel

Isabel of the lily-white hand
    (The wind is sighing in the sedge)
Was walking alone on the brown sea sand
    (By the water's edge, the water's edge)

There she met three sailors blithe and strong
    (The wind is sighing in the sedge)
And the youngest sang a wonderful song
    (By the water's edge, the water's edge)

And the maiden listened and listened long
    (The wind is sighing in the sedge)
"Fain would I learn thy wonderful song."
    (By the water's edge, the water's edge)

"My bark rides yonder on the sea
    (The wind is sighing in the sedge)
So come and I will teach it thee."
    (By the water's edge, the water's edge)

But when they had reached the sailor's bark
    (The wind is sighing in the sedge)
The maiden's sunny brows grew dark
    (By the water's edge, the water's edge)

"What evil chance hath happened thee
    (The wind is sighing in the sedge)

That tears in those blue eyes I see?"
      (By the water's edge, the water's edge)

"The ring my mother gave to me
      (The wind is sighing in the sedge)
Is fallen into the deep blue sea."
      (By the water's edge, the water's edge)

"Oh! dry thine eyes, my lady fair,
      (The wind is sighing in the sedge)
And I will dive and find it there."
      (By the water's edge, the water's edge)

He dived once into the deep blue sea —
      (The wind is sighing in the sedge)
Never a ring to the top brought he
      (By the water's edge, the water's edge)

He dived twice into the deep blue sea —
      (The wind is sighing in the sedge)
And the ring it flashed right gallantly
      (By the water's edge, the water's edge)

He dived thrice into the deep blue sea —
      (The wind is sighing in the sedge)
Never again to the top came he
      (By the water's edge, the water's edge)

Translated from the French
by GEORGE LANIGAN

Cecilia

Although my father's only child,
He sent me o'er the ocean wild.
>   *Sautez, mignonne Cecilia,*
>   *Ah! Ah! Cecilia!*

Over the seas and far away
Borne by a sailor bold and gay.
>   *Sautez &c.*

Borne by a sailor bold and gay.
Who fell in love with me each day.

He fell in love with me each day:
"Ah, Sweet! one little kiss I pray."

"One little kiss for all my care."
"Alas! alas! I'd never dare."

"For if I did," she whispered low,
"My cruel father'd surely know.

"And should he know your love for me,
 A sorely punished maid I'd be."

"Now, foolish maid, we're far away,
 How could your father know, I pray?"

"How could my father know, you say?
 He'd hear it from the wood doves grey."

"But even though the doves might sing,
 He'd never know the tale they bring."

"He would not understand, think you?
 They speak good French — and Latin too."

"Now may his evil neck be wrung
 Who taught the doves the Latin tongue!"
>   *Sautez, mignonne Cecilia,*
>   *Ah! Ah! Cecilia!*

*Translated from the French
by* WILLIAM MCLENNAN

Lullaby

Sleep   sleep beneath the old wind's eye
The wind that's blowing rock-a-bye
To little orioles on their bough
Your dog and duck are sleeping now
An old enchanted spinning-wheel
Has made its home inside your ear
And dreams of spinning in its sleep
From golden sheep
                    sleep . . .

Sleep safe beneath the city's eye
Beyond the field the stricken sky
The night is falling dark and old
The wolf's away   the wind's a-cold
A train goes by   the owl's awake
A mighty ship is under weigh
And leaves behind her on the deep
A port asleep
                    sleep . . .

Sleep tight beneath the eye of Time
An hour   an age   another hour
And every second has its rhyme
The bee is dreaming of the flower
So dream well for the clock's awake
The sweets of Time await the bee
The bear in the wood has gone to sleep
And the snow is deep
                    sleep . . .

GILLES VIGNEAULT
Translated from the French
by JOHN GLASSCO

The Juniper Tree

Meet me my love, meet me my love
By the low branching juniper tree
O I will meet you there my love
If no harm come to me
If no harm come to me

Blue burns the cone, blue burns the cone
Of the low branching juniper tree
And there he waited for his love
As the black minutes go by
As the black minutes go by

Bellows in the field a cow, in the field a cow
Bellows loud after its dead calf
As he waited by the juniper tree
And he heard the red fox cough
He heard the red fox cough

Flew through the air, flew through the air an owl
To the low branching juniper tree
As he waited there for his love
And the black minutes went by
And the black minutes went by

The wind blew down, the wind blew down
Into the low branching juniper tree
And all the seeds rattled in the weed
As the wind blew sudden by
As the wind blew sudden by

Then fell the rain, then fell the rain down
Fell cold into black wet sleet
On the low branching juniper tree
And he said why is she late
He said why is she late

I have come to you my love, my love
Waiting by the juniper tree
And he turns to see her standing there
As white as death was she
As white as death was she

Then why are you so long my love, my love
As I waited at the juniper tree?
But now I will kiss your mouth, he said
O never you will, said she
O never you will, said she

WILFRED WATSON

Abracadabra

In the wicked afternoon
When the witch is there
When the night's downsnare
Swoops like a loon
Strafing the air
In the wicked afternoon

In the witty time of day
When the mind's at play
The cat's at call
The guitar off the wall
Wind holds sway
In the witty time of day

Then the witch will walk
Full of witty talk
And the cat will stalk
Tail high as a cock

The guitar in the room
Will fuss and fume
Strumming at the tune
For a wicked afternoon
And out in the park
Wind will unfrock
The autumn trees
And falling leaves
Shiver with shock

And time with his
Weaving, wailing horn
Shivers my timbers
Shatters my corn:
Little boy blue
Blows a blue tune
On a wicked afternoon.

DOROTHY LIVESAY

The Strangers

Early this morning,
  About the break of day,
Hoofbeats came clashing
  Along the narrow way —

And I looked from my window
  And saw in the square
Four white unicorns
  Stepping pair by pair.

Dappled and clouded,
  So daintily they trod
On small hoofs of ivory
  Silver-shod.

Tameless but gentle,
  Wondering yet wise,
They stared from their silver-lashed
  Sea-blue eyes.

The street was empty
  And blind with dawn —
The shutters were fastened,
  The bolts were drawn,

And sleepers half-rousing
  Said with a sigh,
"There goes the milk,"
  As the hoofs went by!

AUDREY ALEXANDRA BROWN

Flight of the Roller-Coaster

Once more around should do it, the man confided . . .

And sure enough, when the roller-coaster reached the peak
Of the giant curve above me — screech of its wheels
Almost drowned by the shriller cries of the riders —

Instead of the dip and plunge with its landslide of screams
It rose in the air like a movieland magic carpet, some
      wonderful bird,

And without fuss or fanfare swooped slowly across the
      amusement park,
Over Spook's Castle, ice-cream booths, shooting-gallery;
      and losing no height

Made the last yards above the beach, where the cucumber-cool
Brakeman in the last seat saluted
A lady about to change from her bathing-suit.

Then, as many witnesses duly reported, headed leisurely
      over the water,
Disappearing mysteriously all too soon behind a low-lying
      flight of clouds.

RAYMOND SOUSTER

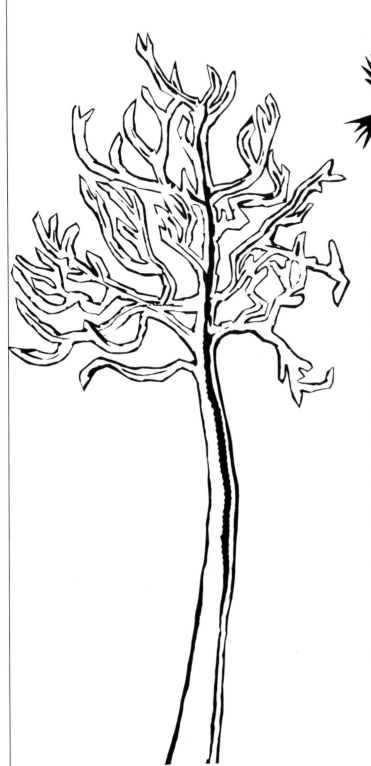

Flight

All day long the clouds go by,
Early winter clouds, not high;
Wide as charity they range,
Restless, regimented, strange.

From my neighbourhood a crow
Takes it in his head to go
Somewhere else he knows about;
Melancholy bird, no doubt.

Up he rises from a tree
Where a yellow leaf or three
Still hang on for hanging's sake,
Tug their yellow stems and shake.

Caw! he cries, as though he knew
Something worth his while to do
In an empty tree elsewhere;
Flap! he takes his blackness there.

Me too! I would like to fly
Somewhere else beneath the sky,
Happy though my choice may be
Empty tree for empty tree.

GEORGE JOHNSTON

I, Icarus

There was a time when I could fly. I swear it.
Perhaps, if I think hard for a moment, I can even tell you
    the year.
My room was on the ground floor at the rear of the house.
My bed faced a window.
Night after night I lay on my bed and willed myself to fly.
It was hard work, I can tell you.
Sometimes I lay perfectly still for an hour before I felt
    my body rising from the bed.
I rose slowly, slowly until I floated three or four feet
    above the floor.
Then, with a kind of swimming motion, I propelled myself
    toward the window.
Outside, I rose higher and higher, above the pasture fence,
    above the clothesline, above the dark, haunted trees
    beyond the pasture.
And, all the time, I heard the music of flutes.
It seemed the wind made this music.
And sometimes there were voices singing.

ALDEN NOWLAN

The Diver

I would like to dive
Down
Into this still pool
Where the rocks at the bottom are safely deep,

Into the green
Of the water seen from within,
A strange light
Streaming past my eyes—

Things hostile;
You cannot stay here, they seem to say;
The rocks, slime-covered, the undulating
Fronds of weeds—

And drift slowly
Among the cooler zones;
Then, upward turning,
Break from the green glimmer

Into the light,
White and ordinary of the day,
And the mild air,
With the breeze and the comfortable shore.

W. W. E. ROSS

Noah

They gathered around and told him not to do it,
They formed a committee and tried to take control,
They cancelled his building permit and they stole
His plans. I sometimes wonder he got through it.
He told them wrath was coming, they would rue it,
He begged them to believe the tides would roll,
He offered them passage to his destined goal,
A new world. They were finished and he knew it.
All to no end.
         And then the rain began.
A spatter at first that barely wet the soil,
Then showers, quick rivulets lacing the town,
Then deluge universal. The old man
Arthritic from his years of scorn and toil
Leaned from the admiral's walk and watched them drown.

ROY DANIELLS

The Shark

He seemed to know the harbour,
So leisurely he swam;
His fin,
Like a piece of sheet-iron,
Three-cornered,
And with knife-edge,
Stirred not a bubble
As it moved
With its base-line on the water.

His body was tubular
And tapered
And smoke-blue,
And as he passed the wharf
He turned,
And snapped at a flat-fish
That was dead and floating.
And I saw the flash of a white throat,
And a double row of white teeth,
And eyes of metallic grey,
Hard and narrow and slit.

Then out of the harbour,
With that three-cornered fin,
Shearing without a bubble the water
Lithely,
Leisurely,
He swam —
That strange fish,
Tubular, tapered, smoke-blue,
Part vulture, part wolf,
Part neither — for his blood was cold.

E. J. PRATT

The Way of Cape Race

Lion-hunger, tiger-leap!
The waves are bred no other way;
It was their way when the Norseman came,
It was the same in Cabot's day:
A thousand years will come again,
When a thousand years have passed away —
Galleon, frigate, liner, plane,
The muster of the slain.

They have placed the light, fog-horn and bell
Along the shore: the wardens keep
Their posts — they do not quell
The roar; they shorten not the leap.
The waves still ring the knell
Of ships that pass at night,
Of dreadnought and of cockle-shell:
They do not heed the light,
The fog-horn and the bell —
Lion-hunger, tiger-leap!

E. J. PRATT

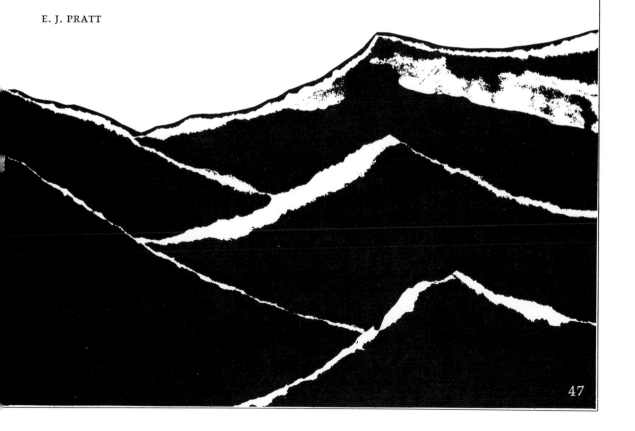

THE GREAT LAKES SUITE

Bodies of Fresh Water

I

Lake Superior

I am Lake Superior
Cold and grey.
I have no superior;
All other lakes
Haven't got what it takes;
All are inferior.
I am Lake Superior
Cold and grey.
I am so cold
That because I chill them
The girls of Fort William
Can't swim in me.
I am so deep
That when people drown in me
Their relatives weep
For they'll never find them.
In me swims the fearsome
Great big sturgeon.
My shores are made of iron
Lined with tough, wizened trees.
No knife of a surgeon
Is sharper than these
Waves of mine
That glitter and shine
In the light of the Moon, my mother
In the light of the Sun, my grandmother.

II
Lake Michigan

For I'm a michigoose
And he's a michigander: OLD VAUDEVILLE SONG

By the shores of Lake Michigan,
Chicago sits
Filled with jawbreakers
Filled with lawbreakers.

By the shores of Lake Michigan,
Lives the Michigander,
Lives the Michigoose.
Very silly people they
For they had the nerve to say
When they used to visit us
In the days of yore,
"That the Yankees,
The Yankees won the war!"

"Bah!" said we
Patriotically.
"How your wits do wander
You Michigoose, you Michigander!"
Right then and there we had a fight
With our cousins from Michigan
Who shortly after went back there again.
And since we won, we knew we were right.

III
Lake Huron

Yoohoo Yoohoo
I'm blue, blue
Lake Huron.
By my shores
In fratricidal wars
Indians killed each other.
At Bayfield
The people stop
To see me slop
Against the pier.
At Grand Bend
The people tend
Instead to
Look at each other.
The Au Sable River and the Maitland
Flow into me.
They think I'm a sea
But haw haw
They're not through yet
For blue and wet
I flow into Lake St Clair
And Lake St Clair into Lake Erie
So very very weary
And Lake Erie into
Lake Ontario
Like a blue grain bag
At which that frowsy hag
Of a city Toronto nibbles.
And then the River St Lawrence!
Whose waters resemble those
Dark barrelled waves that
Drowned the Duke of Clarence.
So haw haw you Maitland River
And you Au Sable one too.
For when you flow into me
You're not at all through.

IV
Lake St Clair

I once knew a bear
Who swam in Lake St Clair
And after the experience
Said, "Hoity Toit
I don't like the way Detroit
Pollutes the air there."
Then after a while
He added with a smile,
"And I don't like the way Windsor
Does, either."

V
Lake Erie

Lake Erie is weary
Of washing the dreary
Crowds of the cities
That line her shores.
Oh, you know
The dirty people of Buffalo
And those in Cleveland
That must leave land
To see what the water's like.
And those that by bike,
Motorcar, bus and screeching train
Come from London in the rain
To Port Stanley where they spend
The day in deciding whether Grand Bend
Might not have been a nicer place to go.
Up and down in thousands
They walk upon Lake Erie's sands.
Those in Cleveland say, "Plainly,"
As they gaze across the waters
Where swim their sons and daughters,
"That distant speck must be Port Stanley."
Those in Port Stanley yawn, "Oh,
That lump in the mist
Over there really must
Be populous Cleveland in Ohio."
But Lake Erie says, " I know
That people say I'm shallow
But you just watch me when I go
With a thump
And a plump
At the Falls of Niagara into Lake Ontario.
When you see that you'll admit
That I am not just a shallow nitwit
But a lake
That takes the cake
For a grand gigantic thunderous tragic exit."

VI
Lake Ontario

Left! Right! march the waves
Towards the sandy shore
Where I stand and motionless
Stare at their blue roar.
Oh, they would stop and listen
And be my blue audience
If I could leap and glisten
More than they, more than they.
But although within me rush
Waves Death cannot deny
I must upon these coasts
Only listen to their cry.
My voice is soft while theirs is loud,
Loud their wavy boasts
That do drown out all reply.
I am one, they are a crowd.
Yet though I'm still and alone
Upon these thin saltless sands,
Thousands only shall hear the waves
Clap their fresh young hands
In lawless blue applause,
Because I held a megaphone
To their blue green blue noise,
Because I made this seashell,
This poem, for your ear,
My dear Monseer,
Of their blue continual hell.

JAMES REANEY

Manerathiak's Song

Kamaoktunga . . . I am afraid and I tremble
When I remember my father and mother
Seeking the wandering game,
Struggling on the empty land
Weakened by hunger. *Eya-ya-ya* . . .

Kamaoktunga . . . I am afraid and I tremble
When I recall their bones
Scattered on the low land,
Broken by prowling beasts,
Swept away by winds. *Eya-ya-ya* . . .

Eskimo chant translated by
RAYMOND DE COCCOLA *and* PAUL KING

The Wind Has Wings

Nunaptigne . . . In our land — *ahe, ahe, ee, ee, iee* —
The wind has wings, winter and summer.
It comes by night and it comes by day,
And children must fear it — *ahe, ahe, ee, ee, iee.*
In our land the nights are long,
And the spirits like to roam in the dark.
I've seen their faces, I've seen their eyes.
They are like ravens, hovering over the dead,
Their dark wings forming long shadows,
And children must fear them — *ahe, ahe, ee, ee, iee.*

Eskimo chant translated by
RAYMOND DE COCCOLA *and* PAUL KING

There's a Fire in the Forest

There's a fire in the forest!
The creatures are fleeing
The flames close behind
With the wind driving onward.
From underbrush up to
The high moving tree-tops
The fire's surging forward.
There's a fire in the forest;
The whole woods are burning.
The whole world is burning!

The creatures are seeking
The safety of streams
Beyond the hot burning.
The creatures are fleeing;
They are labouring, straining
To reach the cool river
They know just beyond them,
To escape the fierce burning,
To reach the cool stream
For which they are yearning.

W. W. E. ROSS

The Rousing Canoe Song

Hide not, hide not,
Deer in lowlands,
Elk in meadows,
Goats on crag-lands.
Hide not brown bear,
Island black bear,
Lynx and cougar,
Mink and beaver.

Safe the martin,
Safe the raccoon,
Now we hunt not
Wolf and cougar,
Brant nor swan
Nor wild geese soaring,
Porpoise, whale
Nor cod nor herring.
Nor bald eagles
From the snow peaks
Curving where the bay is misty.

Lo! We hunt the female otter!
With our spears
We shall surround her.
He who slays her triumphs doubly,
Double prize shall be his portion.

Lo! We hunt the red witch-woman,
Who with magic tricks has harmed us
Even seizing our Great Copper!*

Hide not, hide not,
Game in caverns,
Only hide *thee*, Lost Enchantress!

HERMIA FRASER

*The Copper was a large shield on which
 the names of great chiefs were inscribed.

The Train Dogs

Out of the night and the north;
　　　Savage of breed and of bone,
Shaggy and swift comes the yelping band,
Freighters of fur from the voiceless land
　　　That sleeps in the Arctic zone.

Laden with skins from the north,
　　　Beaver and bear and raccoon,
Marten and mink from the polar belts,
Otter and ermine and sable pelts —
　　　The spoils of the hunter's moon.

Out of the night and the north,
　　　Sinewy, fearless and fleet,
Urging the pack through the pathless snow,
The Indian driver, calling low,
　　　Follows with moccasined feet.

Ships of the night and the north,
　　　Freighters on prairies and plains,
Carrying cargoes from field and flood
They scent the trail through their wild red blood,
　　　The wolfish blood in their veins.

PAULINE JOHNSON

Gluskap's Hound

They slew a god in a valley
    That faces the wooded west:
They held him down in their anger,
    With a mountain across his breast:
And all night through, and all night long,
    His hound will take no rest.

From low woods black as sorrow,
    That marshal along the lake,
A cry breaks out on the stillness
    As if the dead would wake —
The cry of Gluskap's hound, who hunts
    No more for the hunting's sake,

But follows the sides of the valley,
    And the old, familiar trail,
With his nose to the ground, and his eyes
    Red lights in the cedar swale,
All night long and all night through,
    'Til the heavy east grows pale.

Some say he foreheralds tempest,
    Outrunning the wind in the air . . .
When willows are flying yellow
    And alders are wet and bare
He runs with no joy in the running,
    Giving tongue to his mad despair.

Another stick on the fire!
    The shadows are creeping near!
Something runs in the thicket
    The spruces droop to hear!
The black hound running in fierce despair,
    With his grief of a thousand year.

T. G. ROBERTS

From
The Forsaken

Once in the winter
Out on a lake
In the heart of the north-land,
Far from the Fort
And far from the hunters,
A Chippewa woman
With her sick baby,
Crouched in the last hours
Of a great storm.
Frozen and hungry,
She fished through the ice
With a line of the twisted
Bark of the cedar,
And a rabbit-bone hook
Polished and barbed;

Fished with the bare hook
All through the wild day,
Fished and caught nothing;
While the young chieftain
Tugged at her breasts,
Or slept in the lacings
Of the warm *tikanagan*.
All the lake-surface
Streamed with the hissing
Of millions of iceflakes
Hurled by the wind;
Behind her the round
Of a lonely island
Roared like a fire
With the voice of the storm
In the deeps of the cedars.
Valiant, unshaken,
She took of her own flesh,
Baited the fish-hook,
Drew in a grey-trout,
Drew in his fellows,
Heaped them beside her,
Dead in the snow.
Valiant, unshaken,
She faced the long distance,
Wolf-haunted and lonely,
Sure of her goal
And the life of her dear one:
Tramped for two days,
On the third in the morning,
Saw the strong bulk
Of the Fort by the river,
Saw the wood-smoke
Hang soft in the spruces,
Heard the keen yelp
Of the ravenous huskies
Fighting for whitefish:
Then she had rest.

DUNCAN CAMPBELL SCOTT

Eskimo Chant

There is joy in
Feeling the warmth
Come to the great world
And seeing the sun
Follow its old footprints
In the summer night.

There is fear in
Feeling the cold
Come to the great world
And seeing the moon
— Now new moon, now full moon —
Follow its old footprints
In the winter night.

Translated by KNUD RASMUSSEN

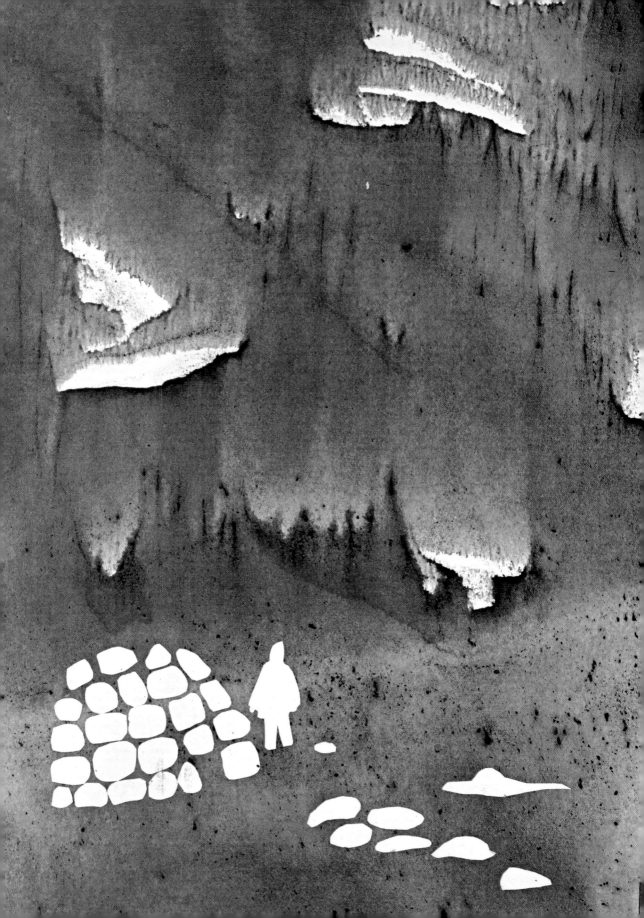

Frost

The frost moved up the window-pane
Against the sun's advance,
In line and pattern weaving there
Rich scenes of old romance —
Armies on the Russian snows,
Cockade, sword, and lance.

It spun a web more magical,
Each moment creeping higher,
For marble cities crowned the hills
With turret, fane and spire,
Till when it struck the flaming sash,
The Kremlin was on fire.

E. J. PRATT

Frost

frost on my window
night comes creaking up

close
stars crowd

GEORGE JOHNSTON

The Ice King

Where the world is grey and lone
Sits the Ice King on his throne —

Passionless, austere, afar,
Underneath the Polar Star.

Over all his splendid plains
An eternal stillness reigns.

Silent creatures of the North,
White and strange and fierce, steal forth:

Soft-foot beasts from frozen lair,
Noiseless birds that wing the air,

Souls of seamen dead, who lie
Stark beneath the pale north sky;

Shapes to living eye unknown,
Wild and shy, come round the throne

Where the Ice King sits in view
To receive their homage due.

But the Ice King's quiet eyes,
Calm, implacable, and wise,

Gaze beyond the silent throng,
With a steadfast look and long,

Down to where the summer streams
Murmur in their golden dreams;

Where the sky is rich and deep,
Where warm stars bring down warm sleep,

Where the days are, every one,
Clad with warmth and crowned with sun.

And the longing gods may feel
Stirs within his heart of steel,

And he yearns far forth to go
From his land of ice and snow.

But forever, grey and lone,
Sits the Ice King on his throne —

Passionless, austere, afar,
Underneath the Polar Star.

A. B. DEMILLE

Ice

When Winter scourged the meadow and the hill
And in the withered leafage worked his will,
The water shrank, and shuddered, and stood still, —
Then built himself a magic house of glass,
Irised with memories of flowers and grass,
Wherein to sit and watch the fury pass.

CHARLES G. D. ROBERTS

All Animals Like Me

All animals like me
now get themselves out of the cold
into some kind of lair
cavernous or small,
to curl up more like a ball
than anything else and sleep
an untroubled sleep of snow,
which sifting down endlessly white
and curdled thick as cream
makes the four-poster of a dream.

RAYMOND SOUSTER

70

After Midnight

Outside beneath the clouds
All the trees stand still
While a white picket fence
Goes racing through the yard.

D. G. JONES

The Rapid

All peacefully gliding,
The waters dividing,
The indolent bateau moved slowly along,
The rowers, light-hearted,
From sorrow long parted,
Beguiled the dull moments with laughter and song:
"Hurrah for the Rapid! that merrily, merrily
Gambols and leaps on its tortuous way;
Soon we will enter it, cheerily, cheerily,
Pleased with its freshness, and wet with its spray."

More swiftly careering,
The wild Rapid nearing,
They dash down the stream like a terrified steed;
The surges delight them,
No terrors affright them,
Their voices keep pace with their quickening speed:
"Hurrah for the Rapid! that merrily, merrily
Shivers its arrows against us in play;
Now we have entered it, cheerily, cheerily,
Our spirits as light as its feathery spray."

Fast downward they're dashing,
Each fearless eye flashing,
Though danger awaits them on every side;
Yon rock — see it frowning!
They strike — they are drowning!
But downward they speed with the merciless tide:
No voice cheers the Rapid, that angrily, angrily
Shivers their bark in its maddening play;
Gaily they entered it — heedlessly, recklessly,
Mingling their lives with its treacherous spray!

CHARLES SANGSTER

The Brook in February

A snowy path for squirrel and fox,
        It winds between the wintry firs.
Snow-muffled are its iron rocks,
        And o'er its stillness nothing stirs.

But low, bend low a listening ear!
        Beneath the mask of moveless white
A babbling whisper you shall hear
        Of birds and blossoms, leaves and light.

CHARLES G. D. ROBERTS

The Worm

Don't ask me how he managed
to corkscrew his way
through the King Street Pavement,
I'll leave that to you.

All I know is
there he was,
circling, uncoiling
his shining three inches,
wiggling all ten toes
as the warm rain fell
in that dark morning street
of early April.

RAYMOND SOUSTER

Eclipse

I looked the sun straight in the eye.
He put on dark glasses.

F. R. SCOTT

At St Jerome

Among the hills of St Jerome,
Though woods are thick and winds are bleak
I would not fear to make my home.

White lilies blow amid the foam
Of waterfalls that outlets seek
Among the hills of St Jerome.

With blueberries and honeycomb,
At Whitefish Lake or St Monique
I would not fear to make my home,

Nor fear to sleep, beneath the dome
Of arching trees with creatures sleek,
Among the hills of St Jerome;

My bed the bracken — book, the tome
Scripted for me on rocky peak,
I would not fear to make my home

Where the Black Mountain grisly gnome
Might nightly wake me with his shriek!
Among the hills of St Jerome
I would not fear to make my home.

FRANCES HARRISON

Psalm of the Fruitful Field:

A field in sunshine is a field
On which God's signature is sealed;
When clouds above the meadow go,
The heart knows peace; the birds fly low.
O field at dusk! O field at dawn!
O golden hay in the golden sun!
O field of golden fireflies
Bringing to earth the starry skies!
You touch the mind with many a gem;
Dewdrops upon the sun's laced hem;
Young dandelions with coronets;
Old ones with beards; pale violets
Sleeping on moss, like princesses;
Sweet clover, purple, odorous;
Fat bees that drowse themselves to sleep
In honey-pots that daisies keep;
Birds in the hedge; and in the ditch
Strawberries growing plump and rich.

Who clamours for a witch's brew
Potioned from hellebore and rue;
Or pagan imps of fairy band,
When merely field and meadowland
Can teach a lad that there are things
That set upon his shoulders wings?
Even a cow that lolls its tongue
Over a buttercup, swells song
In any but a devil's lung.
Even a sheep which rolls in grass
Is happier than lad or lass,
Who treads on stones in streets of brass.
Who does not love a field lacks wit,
And he were better under it!
And as for me let paradise
Set me in fields with sunny skies.
And grant my soul in after days
In clovered meadowlands to graze.

A. M. KLEIN

Windshield Wipers

Windshield wipers
Wipe away the rain,
Please bring the sunshine
Back again.

Windshield wipers
Clean our car,
The fields are green
And we're travelling far.

My father's coat is warm.
My mother's lap is deep.
Windshield wipers,
Carry me to sleep.

And when I wake,
The sun will be
A golden home
Surrounding me;

But if that rain
Gets worse instead,
I want to sleep
Till I'm in my bed.

Windshield wipers
Wipe away the rain,
Please bring the sunshine
Back again.

DENNIS LEE

The Clock Tower

When they pull my clock tower down
I will no longer walk this town.

At night her lucent face is seen
Homely and bright as margarine,

And when I wake when I should sleep
Sounds her Ding Bong sweet
And heart-sticking as the Knife-Man's cry
When his squeaking cart goes by.

Children, chickens,
Matrons with baskets, old men with sticks, all stop
to gawk at my clock;

The shock-headed with the frost
Kid who sells papers, the popcorn man
Buttery-knuckled, the shifter of ashcans,

Firebugs, tire-stealers, track-fixers for the TTC,
Somnambulists, commune with me —

And we all move and love
To the grace of her sweet face.

COLLEEN THIBAUDEAU

Song for Naomi

Who is that in the tall grasses singing
By herself, near the water?
I can not see her
But can it be her
Than whom the grasses so tall
Are taller,
My daughter,
My lovely daughter?

Who is that in the tall grasses running
Beside her, near the water?
She can not see there
Time that pursued her
In the deep grasses so fast
And faster
And caught her,
My foolish daughter.

What is the wind in the fair grass saying
Like a verse, near the water?
Saviours that over
All things have power
Make Time himself grow kind
And kinder
That sought her,
My little daughter.

Who is that at the close of the summer
Near the deep lake? Who wrought her
Comely and slender?
Time but attends and befriends her
Than whom the grasses though tall
Are not taller,
My daughter,
My gentle daughter.

IRVING LAYTON

80

Song

The sun is mine
And the trees are mine
The light breeze is mine
And the birds that inhabit the air
are mine
Their voices upon the wind
are in my ear

ROBERT HOGG

81

Once Upon a Great Holiday

I remember or remember hearing
Stories that began
"Once upon a great holiday
Everyone with legs to run
Raced to the sea, rejoicing."

It may have been harvest Sunday
Or the first Monday in July
Or rockets rising for young Albert's queen.
Nobody knows. But the postman says
It was only one of those fly-by-days
That never come back again.

My brother counted twenty suns
And a swarm of stars in the east,
A cousin swears the west was full of moons;
My father whistled and my mother sang
And my father carried my sister
Down to the sea in his arms.

So one sleep every year I dream
The end of Ramadhan
Or some high holy day
When fathers whistle and mothers sing
And every child is fair of face
And sticks and stones are loving and giving
And sun and moon embrace.

A unicorn runs on this fly-by-day,
Whiter than milk on the grass, so whit

ANNE WILKINSON

Balloon

a s
big as
ball as round
as sun . . . I tug
and pull you when
you run and when
wind blows I
say polite
ly
H
O
L
D
M
E
T
I
G
H
T
L
Y.

COLLEEN THIBAUDEAU

Poem in June

A breeze wipes creases off my forehead
and my trees lean into summer,
putting on for dresses,
day-weave,
ray-weave, sap's green nakedness.

Hushtime of the singers;
wing-time, worm-time
for the squab with its crooked neck and purse-wide beak.
(On wave-blown alfalfa, a hawk-shadow's coasting.)

As a sail fills and bounds with its business of wind,
my trees lean into summer.

MILTON ACORN

June

Now is the ox-eyed daisy out
   Out now everywhere,
The bobolink tips his wings
In the humming blue gold air.

The wild rose opens simple eyes
   In a green briar face,
The mourning dove beats a drum
In a drowsy shady place.

The strawberry like a wren's heart
Shows beneath three green leaves,
The garter snake leaves behind
One of his silver glittering crystal annual sleeves.

JAMES REANEY

85

The Piano

I sit on the edge
of the dining room, almost
in the living room where my parents,
my grandmother, & the visitors
sit knee to knee along the chesterfield & in
the easy chairs. The room is full, & my feet
do not touch the floor, barely
reach the rail across the front
of my seat. "Of course
you will want Bobby to play."—words
that jump out from the clatter
of teacups & illnesses. The piano
is huge, unforgettable.
It takes up the whole end wall
of the living room, faces me down
a short corridor of plump
knees, balanced saucers, hitched
trousers. "Well when is
Bob going to play?"
one of them asks. My dad says,
"Come on, boy, they'd like you
to play for them," & clears
a plate of cake
from the piano bench. I walk between
the knees & sit down
where the cake was, switch on
the fluorescent light
above the music. Right at the first notes
the conversation returns to long tales
of weddings, relatives bombed out again
in England, someone's mongoloid
baby. & there I am at the piano,

with no one listening or even
going to listen
unless I hit sour notes, or stumble
to a false ending.
I finish.
Instantly they are back to me. "What a nice
touch he has," someone interrupts
herself to say.
"It's the hands," says another,
"It's always the hands, you can tell
by the hands." & so I get up
& hide my fists
in my hands.

FRANK DAVEY

From **A Bestiary of the Garden
for Children Who Should Know Better**

on & between the blades of grass
the shining Ant-hoards flow and pass
from morning dew to evening mist
beloved of the fabulist:
    these virtue-hoarders, honey-misers
    timesavers, grasshopper-despisers
    late-to-bedders, early-risers
i think they need some tranquillizers

a somnolent & furry heap
the Rabbit crouches half asleep
upon
the lawn
& one by one the children creep
in running shoes, without a peep
with lots of strings & things to keep
a Rabbit, once it's caught, quite cheap
but woe, alas, it is to weep
just as they reach the sleepy heap
waking & like eugene the jeep
the Rabbit with a sudden leap

the Slug is like a shell-less snail
it has a horned head & a tail

now snails in fishtanks do not fright me
& cooked in garlic, they delight me

why then should that poor shell-less Slug
provoke but one reaction: ug?

the Wasp's a kind of buzzing cousin
of the bee, one of a dozen
of the stingers, striped & yellow
a fussy & officious fellow
can't tell which one? oh, never mind
if you sit on one you'll find
they're all the same to your behind

Zephyr, the lovely wind of summer
rises like a bird, a sea-skimmer
at the twilights, when airs
sweeten & change; flowers
are all moon-faces in the dusk
& there is nothing more to ask
of daybreak than the gift of being
alive under the wing of evening

PHYLLIS GOTLIEB

How and When and Where and Why

How and when and where and why
stars and sun and moon and sky

canals and craters, dunghills, dunes
tell me what's beyond the moons?

beyond the moons the sands are deep
they spread through all the purple skies
in them are Giants who never sleep
but watch the world with burning eyes

they're just like us, with sharper claws
huger pincers, fiercer jaws
and if they catch you—goodbye head!
goodbye little crystal bed!

so wrap your feelers round your feet
fold your thorax nice and neat
the sun is high, the hour is late
now it's time to estivate

> I lay me in my quartzy pool
> I pray the gods to keep it cool
> to keep off demons far and near
> and wake me when the winter's here
> to dance with joy on all my legs
> and live to lay a thousand eggs

PHYLLIS GOTLIEB

A Spider Danced a Cosy Jig

A spider danced a cosy jig
Upon a frail trapeze;
And from a far-off clover field
An ant was heard to sneeze.

And kings that day were wise and just,
And stones began to bleed;
A dead man rose to tell a tale,
A bigot changed his creed.

The stableboy forgot his pride,
The queen confessed an itch;
And lo! more wonderful than all,
The poor man blessed the rich.

IRVING LAYTON

91

Elephant

In his travels, the elephant
    tramps on many planted flower-beds,
and many's the child's plaything
    he crushes under-foot.

His prisonwall-thick flesh
    encloses living blood.

His heavy trunk weaves and bends
    high in the air.

The echoings of his trumpet
    can be heard for many miles.

Stars shine on his back.

DAVID MC FADDEN

Black Bear

Sweet-mouth, honey-paws, hairy one!
you don't prowl much in the history books
but you sure figure when choker-men, donkey-men, shanty-men
gather,
or pulp-savages, or top-riggers.
"I've seen me go up a tree so fast with one of them after me
I only had time to loosen my belt and give him my pants
or I'd have been done for."
"When I came into the cook-house I knew there was something there.
And was there ever! A great big black bear.
He chased me round and round the table till I hauled off and hit the dinner gong.
That shook him! He was out the door like a bat out of hell."
If only you could hear us talk, you would know how we love you
sweet-mouth, honey-paws, hairy one!

Cousin, comrade, and jester,
so like us you pad along jocularly
looking for garbage and honey, and not leaving much trace,
dozing off (for a whole season — as who wouldn't want to?)
then when you waken, perhaps a little too devil-may-care,
not knowing your own strength, ready to carry a joke a little too far
creature of moods, old man, young man, child,
sitting in a meadow eating blue-berries by the bushful.

Don't you know how much we love you?
Old man, curled up in your lair? So come out and be killed, old man!
Sweet-mouth, honey paws, hairy one!

DOUGLAS LEPAN

94

One Step from an Old Dance

Will the weasel lie down with the snowshoe hare?
In the calm and peaceable kingdom?
Will the wolverine cease to rend and tear
In the calm and peaceable kingdom?
Will the beasts of burden not have to bear?
Will the weasel lie down with the snowshoe hare?
Will the children feed grass to the grizzly bear
In the calm and peaceable kingdom?

Oh the wolverine will cease to tear
In the calm and peaceable kingdom,
The rattlesnake rattle praise and prayer
In the calm and peaceable kingdom.
Oh the wolves will wear smiles like children wear,
The wolverine will cease to tear
While the hawk and the squirrel are dancing there
In the calm and peaceable kingdom.

DAVID HELWIG

IN THE CALM AND

The Huntress

The spider works across the wall
    And lo, she droppeth down!
The weather gathers in to fall
    And turns the garden brown.

Poor little spider, late astir,
    Disconsolate she weaves;
No creature comes to nourish her,
    Wherefore she hunts and grieves.

GEORGE JOHNSTON

The Crow

By the wave rising, by the wave breaking
high to low;
by the wave riding the air, sweeping the high air low
in a white foam, in a suds,
there
like a church-warden, like a stiff
turn-the-eye-inward old man
in a cut-away, in the mist
stands
the crow.

P. K. PAGE

The Rattlesnake

An ominous length uncoiling and thin,
A sliver of Satan annoyed by the din
Of six berry-pickers, bare-legged and intent
On stripping red treasure like rubies from Ghent.
He moved without motion, he hissed without noise—
A sombre dark ribbon that laughter destroys;
He eyed them unblinking from planets unknown,
As alien as Saturn, immobile as stone.
With almost forbearance, he watched them retreat—
A creature of deserts, and mountains and heat;
No hint of expression, no trace of regret, no human emotion
     a bland baronet
Secure in his duchy, remote and austere,
Ubiquitous, marvellous grass privateer.

ALFRED PURDY

Indian Summer

Along the line of smoky hills
  The crimson forest stands,
And all the day the blue-jay calls
  Throughout the autumn lands.

Now by the brook the maple leans
  With all his glory spread,
And all the sumachs on the hills
  Have turned their green to red.

Now by great marshes wrapt in mist,
  Or past some river's mouth,
Throughout the long, still autumn day
  Wild birds are flying south.

WILFRED CAMPBELL

Steeds

I have two dashing, prancing steeds,
Buttercup and Dairy Queen,
What for spirit, what for speed,
Matches this amazing team?
One is roan and one is plaid,
One a mare, and one a lad,
One a pacer, one a trotter,
One a son, and one a daughter:
When they're fastened side by side,
Yoked together in the traces,
Joyfully prepare to ride
O'er the big and open spaces;
Whoopee! Swift across the stubble,
Over boulders, banks and rubble,
Up the hill and down the glen,
Cross the country — back again,
Through the fence and greenhouse go,
Pumpkin garden — to and fro,
Pounding, puffing, like a dragon,
Kill the calf and smash the wagon,
Through the hayloft, dust and smother,
In one end and out the other —
Zowie! When their spirit's up!
Dairy Queen and Buttercup!

PAUL HIEBERT

White Cat

I like to go to the stable after supper, —
Remembering fried potatoes and tarts of snow-apple jam —
And watch the men curry the horses,
And feed the pigs, and especially give the butting calves their milk.
When my father has finished milking he will say,
"Now Howard, you'll have to help me carry in these pails.
How will your mother be getting along
All this time without her little man?"
So we go in, and he carries them, but I help.
My father and I don't need the lanterns.
They hang on the wires up high back of the stalls
And we leave them for Ern and Dick.
It seems such a long way to the house in the dark,
But sometimes we talk, and always
There's the White Cat, that has been watching
While my father milked.
In the dark its gallop goes before like air,
Without any noise,
And it thinks we're awfully slow
Coming with the milk.

RAYMOND KNISTER

A Backwards Journey

When I was a child of say, seven
I still had serious attention to give
to everyday objects. The Dutch Cleanser—
which was the kind my mother bought—
in those days came in a round container
of yellow cardboard around which ran
the very busy Dutch Cleanser woman
her face hidden behind her bonnet
holding a yellow Dutch Cleanser can
on which a smaller Dutch Cleanser woman
was holding a smaller Dutch Cleanser can
on which a minute Dutch Cleanser woman
held an imagined Dutch Cleanser can. . . .

This was no game. The woman led me
backwards through the eye of the mind
until she was the smallest point
my thought could hold to. And at that moment
I think I knew that if no one called
and nothing broke the delicate jet
of my attention, that tiny image
could smash the atom of space and time.

P. K. PAGE

103

For the Sisters of the Hôtel Dieu

In pairs,
as if to illustrate their sisterhood,
the sisters pace the hospital garden walks.
In their robes black and white immaculate hoods
they are like birds,
the safe domestic fowl of the House of God.

O biblic birds,
who fluttered to me in my childhood illnesses
— me little, afraid, ill, not of your race, —
the cool wing for my fever, the hovering solace,
the sense of angels —
be thanked, O plumage of paradise, be praised.

A. M. KLEIN

105

Among the Millet

The dew is gleaming in the grass,
    The morning hours are seven,
And I am fain to watch you pass,
    Ye soft white clouds of heaven.

Ye stray and gather, part and fold;
    The wind alone can tame you;
I think of what in time of old
    The poets loved to name you.

They called you sheep, the sky your sward,
    A field without a reaper;
They called the shining sun your lord,
    The shepherd wind your keeper.

Your sweetest poets I will deem
    The men of old for moulding
In simple beauty such a dream,
    And I could lie beholding,

Where daisies in the meadow toss,
    The wind from morn till even,
For ever shepherd you across
    The shining field of heaven.

ARCHIBALD LAMPMAN

Clouds

These clouds are soft fat horses
That draw Weather in his wagon
Who bears in his old hands
Streaked whips and strokes of lightning.
The hooves of his cattle are made
Of limp water, that stamp
Upon the roof during a storm
And fall from dripping eaves;
Yet these hooves have worn away mountains
In their trotting over Earth.
And for manes these clouds
Have the soft and various winds
That still can push
A ship into the sea
And for neighs, the sable thunder.

JAMES REANEY

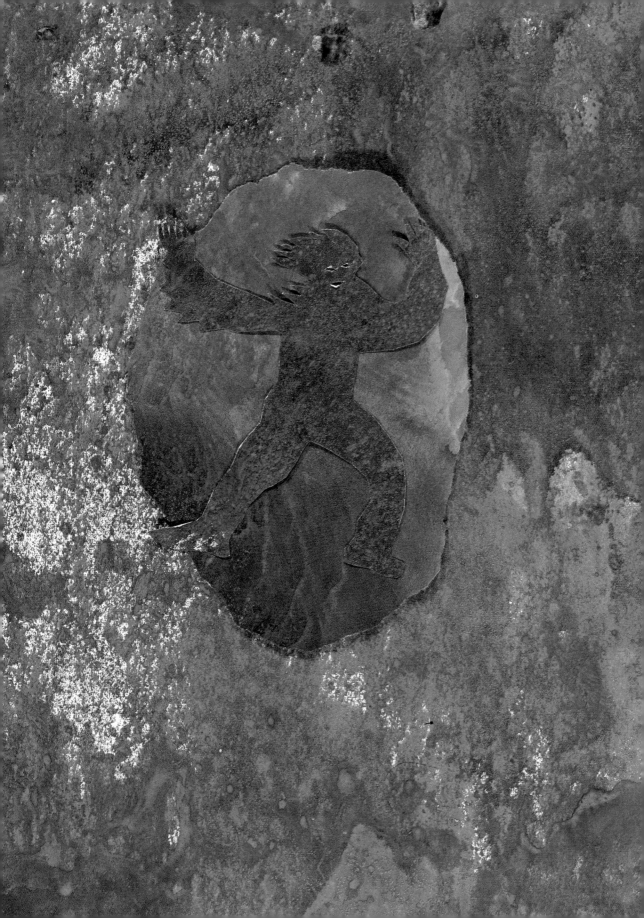

Bye Bye

there's a creaking
in the darkness
a groaning in
the night
there are mutters
in the next
room
oh please
turn on the
light
there are footsteps
in the hallway
there's a
croaking over
there
AAAAAAAAAARRRRGGHHHHHH
a horrid
monster
has got me
by the hair
oh please don't
eat my fingers
oh please
don't bite my
nose
oh please
somebody
save me

GLUMP

SEAN O HUIGIN

Twilight Song

The mountain peaks put on their hoods,
        Good night!
And the long shadows of the woods
Would fain the landscape cover quite;
The timid pigeons homeward fly,
Scared by the whoop owl's eerie cry,
        Whoo-oop! whoo-oop!
As like a fiend he flitteth by;
The ox to stall, the fowl to coop,
The old man to his nightcap warm,
Young men and maids to slumbers light, —
Sweet Mary, keep our souls from harm!
        Good night! good night!

JOHN HUNTER–DUVAR

110

Index of Poets

Acknowledgements

MILTON ACORN. ''Poem in June'' from *Jawbreakers* (1963) by Milton Acorn. HENRY BEISSEL. ''The Boar and the Dromedar'' reprinted by permission of Henry Beissel and his literary agent Arlette Francière. AUDREY ALEXANDRA BROWN. ''The Strangers'' from *Challenge to Time and Death* (1943) by Audrey Alexandra Brown, reprinted by permission of The Macmillan Company of Canada Limited. BLISS CARMAN. ''The Ships of Yule'' from *The Selected Poems of Bliss Carman* ed. by Lorne Pierce, used by permission of The Canadian Publishers, McClelland and Stewart Limited, Toronto. ROY DANIELLS. Used by permission of The Canadian Publishers, McClelland and Stewart Limited, Toronto: ''So They Went Deeper into the Forest...'' from *Deeper Into the Forest* by Roy Daniells; and ''Noah'' from *The Chequered Shade* by Roy Daniells. FRANK DAVEY. Used by permission of the author. RAYMOND DE COCCOLA and PAUL KING. ''Manera-thiak's Song'' and ''The Wind Has Wings'' from *Ayorama* by permission of Oxford University Press Canada. HERMIA FRASER. ''The Rousing Canoe Song'' from *Songs of the Haida* by Hermia Fraser, reprinted by permission of The Ryerson Press, Toronto. PHYLLIS GOTLIEB. All poems used by permission of the author. FRANCES HARRISON. ''At St Jerome'' from *Later Poems* (1928) by S. Frances Harrison, reprinted by permission of The Ryerson Press, Toronto. DAVID HELWIG. ''One Step from an Old Dance'' by David Helwig is reprinted from *Figures in a Landscape* by permission of Oberon Press. PAUL HIEBERT. ''Steeds'' from *Sarah Binks* by Paul Hiebert by permission of Oxford University Press Canada. ROBERT HOGG. Used by permission of the author. GEORGE JOHNSTON. All poems used by permission of the author. D. G. JONES. Used by permission of the author. A. M. KLEIN ''Psalm of the Fruitful Field'' from *Poems* is copyrighted by and used through the courtesy of The Jewish Publication Society of America. Reprinted by permission of The Ryerson Press, Toronto: ''For the Sisters of the Hôtel Dieu'' from *The Rocking Chair* (1948) by A. M. Klein; ''Bandit'' and ''Orders'' from *Hath Not a Jew* (1946) by A. M. Klein. RAYMOND KNISTER. ''White Cat'' from *Collected Poems of Raymond Knister* (1949), reprinted by permission of The Ryerson Press, Toronto. IRVING LAYTON. ''A Spider Danced a Cosy Jig'' and ''Song for Naomi'' from *Collected Poems* © 1965 by Irving Layton, used by permission of The Canadian Publishers, McClelland and Stewart Limited, Toronto. DENNIS LEE. ''Windshield Wipers'' from *Alligator Pie*, copyright © 1974 by Dennis Lee, reprinted by permission of Macmillan of Canada and Houghton Mifflin Company. DOUGLAS LEPAN. ''Black Bear'' from *Something Still to Find* by Douglas LePan, used by permission of the Canadian Publishers, McClelland and Stewart Limited, Toronto. DOROTHY LIVESAY. ''Abracadabra'' from *Selected Poems of Dorothy Livesay* (1957), reprinted by permission of The Ryerson Press, Toronto. DAVID MCFADDEN. Used by permission of the author. JAY MACPHERSON. ''Egg'' from *Poems Twice Told* by permission of Oxford University Press Canada. ALDEN NOWLAN. ''I, Icarus'' from *Bread, Wine and Salt* by Alden·Nowlan © 1967 by Clarke Irwin (1983) Inc. Used by permission. SEAN O HUIGIN. ''Bye Bye'' used by permission of Black Moss Press. P. K. PAGE. Poems used by permission of the author. E. J. PRATT. ''Frost'', ''The Shark'', and ''The Way of Cape Race'' from *Collected Poems of E. J. Pratt* (1958), reprinted by permission of The Macmillan Company of Canada Limited. ALFRED PURDY. Used by permission of the author. KNUD RASMUSSEN. ''Eskimo Chant'' from *Beyond the High Hills: A Book of Eskimo Poems* (1961) by Knud Rasmussen, reprinted by permission of The World Publishing Company, New York. JAMES REANEY. For permission to reprint ''June'', ''The Great Lakes Suite'', and ''Clouds'' from *Poems*, copyright 1972 by James Reaney, thanks are due to the author and New Press. CHARLES G. D. ROBERTS. ''Ice'' and ''The Brook in February'' from *Selected Poems* (1936) by Charles G. D. Roberts, reprinted by permission of The Ryerson Press, Toronto. T. G. ROBERTS. Reprinted by permission of The Ryerson Press, Toronto: ''The Reformed Pirate'' from *Cap and Bells: An Anthology of Light Verse by Canadian Poets* (1936) ed. by John Garvin, ''Gluskap's Hound'' from *The Leather Bottle* (1934) by T. G. Roberts. W. W. E. ROSS. ''The Diver'' and ''There's a Fire in the Forest'' from *Shapes and Sounds* (1968) reprinted by permission of Academic Press Canada. DUNCAN CAMPBELL SCOTT. From ''The Forsaken'' used by permission of John G. Aylen, Ottawa, Canada. F. R. SCOTT. ''Eclipse'' from *Collected Poems* by F. R. Scott used by permission of The Canadian Publisher, McClelland and Stewart Limited, Toronto. Y. Y. SEGAL. ''King Rufus'' used by permission of Sandor Klein. ''Rhymes'' used by permission of Miriam Waddington. ROBERT W. SERVICE. ''The Shooting of Dan McGrew'', from *Collected Poems of Robert Service* (1966), reprinted by permission of Dodd, Mead and Company Inc., New York, The Ryerson Press, Toronto, and Feinman and Krasilovsky, Attorneys. VIRNA SHEARD. ''The Yak'' from *Leaves in the Wind* (1938) by Virna Sheard, reprinted by permission of The Ryerson Press, Toronto. RAYMOND SOUSTER. ''The Worm'' reprinted from *Collected Poems of Raymond Souster* by permission of Oberon Press. Reprinted by permission of The Ryerson Press, Toronto: ''All Animals Like Me'' from *Ten Elephants on Yonge Street* (1965) by Raymond Souster and ''Flight of the Roller-Coaster'' from *The Colour of the Times* (1964) by Raymond Souster. COLLEEN THIBAUDEAU. Both poems used by permission of the author. GILLES VIGNEAULT. ''Lullaby''/''Berceuse'' by Gilles Vigneault, copyright NOUVELLES EDITIONS DE L'ARC. Montreal, Quebec, reprinted in a translation by John Glassco by permission of Gilles Vigneault and William Toye. MIRIAM WADDINGTON. ''Laughter'' from *Driving Home* by permission of Oxford University Press Canada. WILFRED WATSON. ''The Juniper Tree'' from *Friday's Child* reprinted by permission of Faber and Faber Ltd. and Farrar, Straus and Giroux, Inc. ANNE WILKINSON. ''Once Upon a Great Holiday'' from *The Collected Poems of Anne Wilkinson* (1968) ed. by A. J. M. Smith, reprinted by permission of The Macmillan Company of Canada Limited.